I0471157

Dutch Pedestrian Safety Research

Review

PUBLICATION NO. FHWA-RD-99-092

DECEMBER 1999

U.S. Department of Transportation

Federal Highway Administration

Research, Development, and Technology
Turner-Fairbank Highway Research Center
6300Georgetown Pike
McLean, VA 22101-2296

FOREWORD

Creating improved safety and access for pedestrians requires providing safe places for people to walk, as well as implementing traffic control and design measures which allow for safer street crossings. A study entitled "Evaluation of Pedestrian Facilities" involved evaluating various types of pedestrian facilities and traffic control devices, including pedestrian crossing signs, marked versus unmarked crosswalks, countdown pedestrian signals, illuminated pushbuttons, automatic pedestrian detectors, and traffic calming devices such as curb extensions and raised crosswalks. The study provided recommendations for adding sidewalks to new and existing streets and for using marked crosswalks for uncontrolled locations. The "Evaluation of Pedestrian Facilities" also included synthesis reports of both domestic and international pedestrian safety research. There are five international pedestrian safety synthesis reports; this document compiles the most relevant research from the Netherlands.

This synthesis report should be of interest to State and local pedestrian and bicycle coordinators, transportation engineers, planners, and researchers involved in the safety and design of pedestrian facilities within the highway environment.

Michael F. Trentacoste
Director, Office of Safety
Research and Development

NOTICE

This document is disseminated under the sponsorship of the Department of Transportation in the interest of information exchange. The U.S. Government assumes no liability for its contents or use thereof. This report does not constitute a standard, specification, or regulation.

The U.S. Government does not endorse products or manufacturers. Trade and manufacturer's names appear in this report only because they are considered essential to the object of the document.

1. Report No. FHWA-RD-99-092	2. Government Accession No.	3. Recipient's Catalog No.
4. Title and Subtitle Dutch Pedestrian Safety Research Review		5. Report Date
		6. Performing Organization Code
7. Author(s) T. Hummel		8. Performing Organization Report No.
9. Performing Organization Name and Address SWOV Institute for Road Safety Research The Netherlands University of North Carolina Highway Safety Research Center 730 Airport Rd, CB #3430 Chapel Hill, NC 27599-3430		10. Work Unit No. (TRAIS)
		11. Contract or Grant No. DTFH61-92-C-00138
12. Sponsoring Agency Name and Address Federal Highway Administration Turner-Fairbank Highway Research Center 6300 Georgetown Pike McLean, VA 22101-2296		13. Type of Report and Period Covered
		14. Sponsoring Agency Code

15. Supplementary Notes

Prime Contractor: University of North Carolina Highway Safety Research Center
FHWA COTR: Carol Tan Esse

16. Abstract

This report was one in a series of pedestrian safety synthesis reports prepared for the Federal Highway Administration (FHWA) to document pedestrian safety in other countries. Reports are also available for:

United Kingdom (FHWA-RD-99-089)
Canada (FHWA-RD-99-090)
Sweden (FHWA-RD-99-091)
Australia (FHWA-RD-99-093)

This report is a review of recent pedestrian safety research in the Netherlands. It addresses several topics, reporting findings and providing a comprehensive list of references. Topics addressed include:

Pedestrian crossings and traffic calming measures: Here research is reviewed on pedestrian crossings; along with other research pertaining to infrastructure changes in the form of traffic calming.
Children and the elderly: One study shows that children are now less likely to walk to school than in earlier times because of parental concern for their safety. Measures for increasing safety of elderly pedestrians are also presented.
Disabled pedestrians: Discussion is included concerning hardware and infrastructure that perhaps could be made in order to give better consideration to pedestrians with some kind of disability.
Passenger car front-end structure: Discussion is presented as to the role of the car's structural properties as it influences injury severity in a collision with a pedestrian.

17. Key Words: pedestrian safety, pedestrian crossings, traffic calming, disabled pedestrians		18. Distribution Statement

19. Security Classif. (of this report) Unclassified	20. Security Classif. (of this page) Unclassified	21. No. of Pages 37	22. Price

Form DOT F 1700.7 (8-72)   Reproduction of form and completed page is authorized

SI* (MODERN METRIC) CONVERSION FACTORS

APPROXIMATE CONVERSIONS TO SI UNITS

Symbol	When You Know	Multiply by	To Find	Symbol
LENGTH				
in	inches	25.4	millimeters	mm
ft	feet	0.305	meters	m
yd	yards	0.914	meters	m
mi	miles	1.61	kilometers	km
AREA				
in²	square inches	645.2	square millimeters	mm²
ft²	square feet	0.093	square meters	m²
yd²	square yards	0.836	square meters	m²
ac	acres	0.405	hectares	ha
mi²	square miles	2.59	square kilometers	km²
VOLUME				
fl oz	fluid ounces	29.57	milliliters	mL
gal	gallons	3.785	liters	L
ft³	cubic feet	0.028	cubic meters	m³
yd³	cubic yards	0.765	cubic meters	m³
NOTE: Volumes greater than 1000 l shall be shown in m³.				
MASS				
oz	ounces	28.35	grams	g
lb	pounds	0.454	kilograms	kg
T	short tons (2000 lb)	0.907	megagrams (or "metric ton")	Mg (or "t")
TEMPERATURE				
°F	Fahrenheit temperature	5(F-32)/9 or (F-32)/1.8	Celcius temperature	°C
ILLUMINATION				
fc	foot-candles	10.76	lux	lx
fl	foot-Lamberts	3.426	candela/m²	cd/m²
FORCE and PRESSURE or STRESS				
lbf	poundforce	4.45	newtons	N
lbf/in²	poundforce per square inch	6.89	kilopascals	kPa

APPROXIMATE CONVERSIONS FROM SI UNITS

Symbol	When You Know	Multiply by	To Find	Symbol
LENGTH				
mm	millimeters	0.039	inches	in
m	meters	3.28	feet	ft
m	meters	1.09	yards	yd
km	kilometers	0.621	miles	mi
AREA				
mm²	square millimeters	0.0016	square inches	in²
m²	square meters	10.764	square feet	ft²
m²	square meters	1.195	square yards	yd²
ha	hectares	2.47	acres	ac
km²	square kilometers	0.386	square miles	mi²
VOLUME				
mL	milliliters	0.034	fluid ounces	fl oz
L	liters	0.264	gallons	gal
m³	cubic meters	35.71	cubic feet	ft³
m³	cubic meters	1.307	cubic yards	yd³
MASS				
g	grams	0.035	ounces	oz
kg	kilograms	2.202	pounds	lb
Mg (or "t")	megagrams (or "metric ton")	1.103	short tons (2000 lb)	T
TEMPERATURE				
°C	Celcius temperature	1.8C+32	Fahrenheit temperature	°F
ILLUMINATION				
lx	lux	0.0929	foot-candles	fc
cd/m²	candela/m²	0.2919	foot-Lamberts	fl
FORCE and PRESSURE or STRESS				
N	newtons	0.225	poundforce	lbf
kPa	kilopascals	0.145	poundforce per square inch	lbf/in²

(Revised September 1993)

*SI is the symbol for the International System of Units. Appropriate rounding should be made to comply with Section 4 of ASTM E380.

Contents

1. Summary of pedestrian accident experience

This chapter offers a description of general information concerning the development of traffic safety and mobility of pedestrians in the Netherlands. In addition, two studies concerning pedestrian accident experience will be summarized.

In the first study, by R. Methorst (1993), demographic and social trends are determined by means of a survey. These trends can be used as (part of) an explanation of developments in pedestrian safety. According to the author, future developments will not result in an increase in the number of pedestrian accidents, but rather in a limitation of pedestrian mobility and freedom of movement. The second study, by L.T.B. van Kampen (1991), contains an analysis of injury data of both pedestrians and bicyclists. Results of this analysis are used to determine possible means of reducing the severity of injuries. Protective clothing, safety helmets, and car front design are taken into consideration. The subject of car front design is discussed in greater detail in chapter 8, "Car front impact requirements."

The number of pedestrians killed in a traffic accident has decreased strongly in the 1980's. This decrease came to an end in the 1990's. The year 1996 however shows a remarkably positive development.

The stagnation in the decrease can hardly be explained by an increase in pedestrian mobility for this has remained relatively unchanged between 1980 and 1994 (5 to 5.5 billion km (3 to 3.5 billion mi) per year). In recent years the level of pedestrian mobility is somewhat higher.

Estimates of pedestrian mobility are made by CBS (Statistics Netherlands). According to estimates of the Dutch Pedestrians Association, these official pedestrian mobility estimates are too low. Estimates of the Pedestrians Association indicate a pedestrian mobility 8 billion km (5 billion mi) per year (for the year 1994).

Year	Pedestrian fatalities	Pedestrian mobility 10^9 km.	Fatal acc. per 10^9 km traveled
1985	187	5.1	37
1986	216	4.9	44
1987	172	5.0	34
1988	201	4.8	42
1989	190	5.0	38
1990	144	5.0	29
1991	144	5.2	28
1992	152	5.2	29
1993	146	5.2	28
1994	123	5.7	22
1995	142	5.6	25
1996	109	5.5	20

Table 1. *Pedestrian fatalities, pedestrian mobility, and exposure to risk in traffic.*

Mode of transport	deaths/10^9 km.	hosp. injur./10^9 km.
Car/Van	4	39
Truck/Bus	1	6
Motorbike	59	815
Moped	87	2.537
Bicycle	20	545
Pedestrian	22	291
Total	6	90

Table 2. *Exposure to risk in traffic for different transport modes; average over years 1994, 1995, 1996. Deaths and hospitalised injuries per 10^9 km.*

Pedestrians mainly are killed in accidents with cars. Pedestrian safety therefore is also determined by the mobility of motorized traffic. This mobility of motorized traffic is still growing every year.

Children and elderly pedestrians prove to be the most vulnerable. Nearly 50 percent of the total number of killed pedestrians is older than 65 years. Their risk, expressed as the number of deaths per km, is also found to be very high: more than 100 deaths per billion km (62 mi) compared to 27 (17 mi) on average for all age groups (Accident Records Registration Division of the Directorate-General of Public Works, 1980-1997).

Next to the elderly, children 14 and under are the second most vulnerable age group. The number of children killed in a traffic accident has however decreased more than in other age groups (Accident Records Registration Division of the Directorate-General of Public Works, 1980-1997).

It is a known fact that not all traffic accidents are registered. The registration of deaths, however, is known to be complete (or nearly complete). The registration of injury accidents is not. The shown figures are corrected for this under registration.

Age	1990	1991	1992	1993	1994	1995	1996
0 - 5	22	24	25	26	17	25	10
6 - 12	56	45	46	49	48	35	28
13-15	36	29	35	24	39	40	48
16-17	67	80	51	62	41	62	49
18-25	313	305	279	266	283	255	258
26-50	424	377	391	387	421	445	389
51-64	151	149	147	121	160	163	125
> 65	307	272	311	317	289	309	273

Table 3. *Deaths per year for different age groups (for all transport modes).*

1.1. Methorst (1993)

In most countries pedestrian mobility and safety are not considered to be important issues. As a result, the national statistical agencies pay little attention to pedestrian mobility and safety. This in turn leads to under-estimation of present and future pedestrian problems.

Trends in these pedestrian problems are not recognized. Not surprisingly, very few strategies are developed, and no action is undertaken to alleviate the problems.

In this paper, the author tries to break through this vicious circle by improving insight in both the present and future position of the pedestrian, in particular the relation between pedestrian mobility and safety.

The Dutch Pedestrians Association has carried out a survey on the relevance and representativeness of data regarding pedestrian mobility and safety. Demographic and social trends were identified and used as input for a prognosis of pedestrian mobility and safety.

The study was limited to the Netherlands. Much of what is contained here is likely to be relevant to the USA situation, although the higher proportion of travel by passenger car in the U.S. should be kept in mind.

Some of the findings are:

- Approximately 3 percent of the total distance traveled is traveled by foot.
- Approximately 20 percent of the total number of trips is done by foot.
- The average citizen in the Netherlands walks about 1,600 times per year and uses a car only 650 times a year.

In the Netherlands, walking appears to be safe. According to Statistics Netherlands, the number of pedestrians killed or injured in accidents has decreased substantially over the last two decades.

Several studies have shown, however, that the reduction in casualties is largely caused by self-imposed restrictions in mobility by pedestrians.

Because of several factors (travel time budgets, demographic trends, trends in living conditions, educational levels, equal opportunities for women, employment trends, and enlargement of scale) car ownership and car use will increase dramatically in the next 20 years. As a result, pedestrians in the future would have less space for care-free, undisturbed, and safe walking.

The author stresses the need for governmental intertervention to prevent this situation from getting out of hand.

1.2. Van Kampen (1991)

Accidents with cyclists and pedestrians tend to be more serious when motorized traffic is involved. In the project "Safe bicycle and injury prevention" special attention is given to the bicycle, the car front, and the protection of the cyclist.

Part of this study is covered by this report. Research is conducted to answer the question if analysis of registered traffic accidents can lead to judgments of the specific needs of cyclist and pedestrian protection.

Comparing the findings for cyclists and pedestrians, it comes out that both type and severity of the injuries of cyclists and pedestrians show a great deal of similarity. In the group of pedestrians however, the proportion of injury to the legs was significantly greater than in the group of cyclists.

Thirty-four percent of the injured pedestrians suffer injuries to the head and skull; 33 percent of the injured pedestrians suffer injuries to the legs. Among elderly pedestrians, the proportion of injuries to the legs is greater than average, mainly caused by the relatively high proportion of injuries to the upper legs.

Because of the high risk of permanent consequences of these types of injuries, there seems to be an obvious need for measures to protect the legs of pedestrians. In view of the types of injuries, both cyclists and pedestrians should be protected against injuries to the head/skull. The wearing of safety helmets by cyclists could be a good solution. For pedestrians, however, the wearing of a safety helmet does not seem a very obvious measure. Reconstruction of the cars' front end should provide important contributions to the protection of both cyclists and pedestrians.

The proportion of injuries to the legs for pedestrians is similar to the proportion of injuries to the head/skull. It can therefore be concluded that protection of the legs of pedestrians is important. The wearing of protective clothing is not an obvious measure. Reconstruction of the car's front end, especially the bumper and its surroundings, is an important measure.

2.   Pedestrian crossings

This chapter outlines several studies on pedestrian crossings. Attention is paid to both safety aspects of signalized and unsignalized pedestrian crossings, as well as to innovative measures to improve signalized pedestrian crossings. Several studies are mentioned briefly right below and are fully described thereafter.

The first study (Boot, 1987; see Section 3.1) contains an analysis of traffic accidents on signalized and unsignalized crossings in the Netherlands. Results of this analysis are compared with similar data on the safety of Swiss pedestrian crossings. The results show that installation of unsignalized pedestrian crossings does not lead to an improvement of traffic safety. Signalized crossings in situations with high volumes of motorized traffic and pedestrian traffic however proved to have a positive effect on traffic safety.

The second study (Lange, 1996; see Section 3.2) comprises an observational examination of crossing pedestrians. Examined is which factors influence jaywalking at signalized crossings.

Three studies (Levelt, 1994; Janssen & Van der Horst, 1991; Levelt, 1992) concern innovative measures for improvement of signalized crossings. *Pedestrian Opinions on the Alternative Maastricht Crossing* (Levelt, 1994; see Section 3.3) covers a survey of pedestrians using the alternative Maastricht crossing. Traditional signalized pedestrian crossings in the Netherlands consist of red light (standing man) above a green light (walking man), positioned across the street. Before the green light changes to red it flashes for a short period. In the alternative Maastricht crossing, the same traffic light is positioned on the near side of the crossing instead of the opposite side.

In *An Evaluation of Flashing Yellow at Signalized Pedestrian Crossings* (Janssen & Van der Horst, 1991; see Section 3.4), the behavior of crossing pedestrians using the alternative flashing yellow traffic light is observed. In this type of crossing, the red light in the traditional pedestrian light is replaced by a flashing yellow light. Whereas the traditional red light means "forbidden to cross," the flashing yellow light means "there could be conflicting traffic; crossing is at your own risk." In the alternative setting, green light always means "no conflicting traffic," which is not always the case with the traditional pedestrian lights.

The most far reaching alternative is discussed in *The Dutch Experiment With Pussycats* (Levelt, 1992; see Section 3.5). This study consists of an observation and survey of pedestrians using the new type of pedestrian crossing called Pussycats. This type of crossing can be described as an advanced combination of the two alternative crossings described earlier. The pedestrian display consists of a green light (walking man) and a flashing yellow light, and is positioned on the near side of the crossing (the Maastricht position). Further, waiting pedestrians as well as crossing pedestrians are detected and monitored. These technical improvements make it possible to show the pedestrian green light for short periods, cancel unused calls, and adjust the clearance time for slow pedestrians and large groups.

The last study described in this chapter (Carsten et al., 1992; see Section 3.6) involves the development of simulation models that represent the movement of pedestrians around a street network and the safety consequences of the various road crossing flows.

2.1. Boot (1987)

Signalized and unsignalized pedestrian crossings usually are realized to improve traffic safety. Examinations of traffic accidents on crossings in the Netherlands, however, show that the installation of unsignalized crossings doesn't lead to an improvement of traffic safety. In some cases, the number of accidents is found to be increased after the installation of an unsignalized crossing.

In contradiction with these findings, research in Switzerland showed an improvement of traffic safety after the installation of unsignalized pedestrian crossings. These results can, however, be biased by the fact that only crossings in rural municipalities have been studied and that only fatal accidents (which are fortunately very rare) have been studied.

One of the positive aspects in the results of the Swiss research of (unsignalized) pedestrian crossings was the concentration of crossing pedestrians at one location. Given the bad crossing discipline of Dutch pedestrians, it can be doubted if these results can be translated to the situation in the Netherlands.

The installation of signalized crossings in the Netherlands, according to the criteria used, however, proved to have a positive effect on traffic safety. It must be understood that signalized crossings in the Netherlands are only realized when volumes of motorized traffic as well as crossing pedestrians are high.

Recommendations
-   Unsignalized crossings: The author stresses that a revision of the legal status of unsignalized pedestrian crossings is needed. At present, pedestrians only have right of way when they are already on the crossing. To reduce both waiting times and dangerous conflicts, pedestrians waiting to cross should also have priority.

    Installation of unsignalized crossings should only be considered if no more than one traffic lane per direction is crossed. If traffic speeds exceed 50 km/h (31 mi/i), the installation of unsignalized crossings should be advised against.

-   Signalized crossings: Installation of signalized crossings should only be taken in consideration if volumes of both motorized traffic and pedestrian traffic are high.

2.2. De Lange (1986)

This report describes a method to determine the safety and freedom of movement of pedestrians at crossing places. Research has been carried out at crossing places with high volumes of traffic as well as high volumes of crossing pedestrians.

Fifty-three percent of the interviewed pedestrians state that high speeds of traffic approaching the crossing impedes the pedestrians while crossing. The influence of the length of waiting times at signalized crossings on the number of jaywalkers proved to be smaller than assumed. It therefore can be concluded that reduction of waiting time can only have relatively small effects on the number of jaywalking pedestrians.

The type of destination of the pedestrians was found to have no effect on the chance of crossing on red. Age, however, proved to be a significant influence. The percentage of pedestrians of 65 years and older crossing on red light is significantly smaller than the percentage of younger pedestrians crossing on red. In contrast with younger respondents, pedestrians of 65 years and older don't regard waiting times at signalized crossings as a problem.

Furthermore, elderly pedestrians do not adequately judge speeds of approaching traffic. This can be explained by the habit of only crossing on green.

2.3. Levelt (1994)

The alternative layout for pedestrian crossings, the Maastricht design, in which the light is positioned on the near side of the crossing, is under discussion. One of the arguments against introduction of this alternative is the supposed resistance felt by pedestrians — a resistance that has not been expressed so much through complaints lodged with the road planning authorities but rather through several polls held among pedestrians on the street.

The CROW[1] working group for pedestrian engineering facilities wished to know if this perceived resistance should be taken into account in the recommendation.

The CROW asked the SWOV Institute for Road Safety Research to conduct a study among users of the Maastricht crossing to investigate the presence and if so whether such resistance can be overcome through information campaigns.

The SWOV questioned 200 pedestrians at 29 crossings with the Maastricht design, which involved nine locations in two municipalities. First, people were asked to state the characteristic differences, then their preference was asked, and finally a comment about perceived safety was requested. The background to the response in favor of one or other layout was questioned. Subsequently, the opinion about a number of characteristics associated with the new layout was requested.

Some information regarding possible principal advantages was given to the respondent: time won with a short green interval, better visibility for the partially sighted, and loss of the fright response among elderly

---

[1]CROW: Netherlands Centre for Research and Contract Standardisation in Civil and Traffic Engineering.

Figure 1. *Maastricht design crossing.*

when they are confronted by a red light while crossing. Subsequently, the interviewee's preference and safety assessment was once again requested. In this way, it was attempted to obtain an insight into the nature of possible resistances, and it was studied whether information about the advantages of the new design would be able to alleviate resistance.

The first striking result was that less than half of those interviewed were able to cite the actual main distinguishing characteristics: the change in position of the pedestrian light. Exposure to the system did not influence this response. The second, most important result was that there did not seem to be great resistance to the new design. On the contrary, 32 percent preferred this layout, 22 percent preferred the old layout, and 44 percent demonstrated no preference. The safety assessment, which is strongly related to preference, did not favor either of the two systems: 27 percent judged the Maastricht layout safer, whereas 29 percent judged the old layout as safer; 44 percent demonstrated no preference.

In view of previous study results, these outcomes were not anticipated. People who had used the crossing for a period of over a year for at least once a week preferred the new system.

The advantages and disadvantages cited by people with preference for one of the two systems were related both to the characteristics specific to the system and to characteristics that can also be found elsewhere. Relevant advantages quoted in particular were that the light is more visible and that it is more suitable for the elderly and the partially sighted.

Further advantages cited included the presence of a push button to request a green light and the presence of a sound signal. The primary disadvantages mentioned were the lack of an opposite pedestrian light, uncertainty about which point of time the traffic would start to move, and inability to see the light turn red, so the pedestrian is unsure whether (s)he needs to hurry. Those in support of the Maastricht design in general cited more advantages than the opponents were able to cite disadvantages.

When asked about general positive characteristics of the new layout, people confirmed in general that they are given sufficient time to cross in this system, that the partially sighted are better able to see the light, and that the sound signal clearly indicates that the light has switched to green.

People did not agree that they actually are safer while crossing. With regard to negative characteristics, people reiterated [in the main] that they have more crossing time with the old system, that they do not know at which moment the traffic will start to move, and that they are more inclined to cross on red with the new system. They deny that the traffic starts moving as soon as the sound signal stops and that two systems operating in parallel would be confusing. The inconsistency in the remarks: "sufficient time to cross" and "more crossing time with the old system" could largely be explained by the fact that these remarks were given by different respondents. A large number of opinions related to personal preference and the safety assessment.

The information given during the interview and the three above mentioned advantages did not lead to a shift in preference or in the safety assessment. Comparison to previous studies supports the assumption that resistance is primarily seen with a change to the existing situation, while there is less resistance to introduction at locations where the crossing was not yet controlled. It was found that only 35 percent of the pedestrians cross exclusively on green, and that half of those crossing on red press the request button first.

It is recommended that in the process of assessing the Maastricht design the resistance expressed by pedestrians should not be taken into account, and neither should a possible variation in uniformity. Attention is asked — with regard to the installation and information campaigns — for giving pedestrians the option to request green, for sound signals, and for sufficient crossing time, if possible by using detectors for crossing pedestrians. It is again emphasized that unnecessary requests for a green light should be avoided, again through the use of detector systems.

2.4. Janssen & Van der Horst (1991)

The replacement of red by blinking yellow has been investigated on six different pedestrian crossings in Delft with the aim of evaluating the effects on pedestrian behavior. The investigation was performed in 1989 and 1990 by means of a before/after study. Video registration as well as conflict observations on the spot were applied as investigation methods.

Video results included number of pedestrians crossing in the separate phases of the cycle, as well as gaps accepted or rejected by crossing pedestrians. A distinction was made between vulnerable pedestrians (children and elderly people) and the remaining group. The main results are as follows:

- The percentage of pedestrians not crossing in the green phase has, on average, been doubled by the introduction of blinking yellow. As a consequence, average waiting times have been reduced.

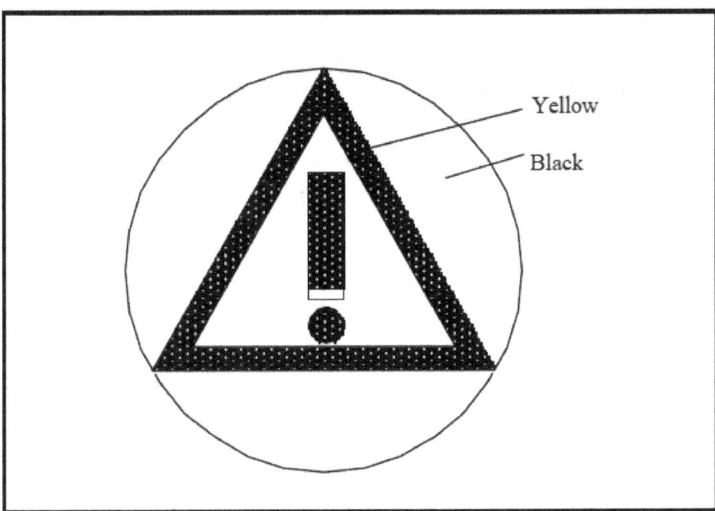

Figure 2. *Flashing yellow pedestrian signal.*

- The size of the so-called "critical gap" when crossing outside green is not affected by the blinking yellow.
- There was no change in the number of conflicts observed when pedestrians crossed outside the green phase after replacement of the red with the blinking yellow.
- There was no indication that the risk while crossing was in some way specifically increased for vulnerable pedestrians.

Figure 3. *Flashing yellow pedestrian signals.*

Figure 4. *Traditional pedestrian signals (green: walking man).*

It was concluded that comfort at the experimental pedestrian crossing had been improved by the introduction of blinking yellow. It was also concluded that crossing during blinking yellow had by itself not become more dangerous than crossing during the red light previously. However, in so far as crossing outside green will, in principle, be more dangerous than crossing during green, the net results for safety could be negative because of the doubling in the number of people crossing outside during blinking yellow.

The suggestion to replace the red light for pedestrians with the blinking yellow would meet the wish of pedestrians not having to wait unnecessarily and make the decision whether to cross themselves.

As a positive side effect, introduction of blinking yellow would mean an unequivocal relation between the sign of the pedestrian light and the possible arrival of conflicting traffic (the traditional green pedestrian light doesn't necessarily mean that there isn't any conflicting traffic). Furthermore, the existing level of jaywalking would lapse, thus reducing the possible blurring of moral standards in traffic.

Recommendations
1. On midblock crossings, the amount of jaywalkers is so low and the waiting times are so short that there seems to be no reason to replace the red light with a blinking yellow light.
2. At crossings of major traffic streams, there is a relatively great willingness to wait at a red light. The replacement of red with blinking yellow in this situation is unnecessary here too.
3. On crossings at minor traffic streams (parallel with major traffic streams), the willingness to wait at a red light is very low. In situations like this, red pedestrian lights could be replaced with blinking yellow lights.

2.5. Levelt (1992)

This report is the Dutch part of an international (French, British, and Dutch) evaluation study of new pedestrian crossing facilities, merged under the name "Pussycats."

Dutch pedestrian signals consist of a red light (standing man) above a green one (walking man) positioned across the street. Before the green light changes to red, it flashes for a short period. Pedestrians may still start to cross during flashing green. Red means: "if you are on the crossing, move to the curb as quickly as possible" and otherwise "do not cross."

In the Netherlands new traffic regulations went into effect in 1991. These regulations (RVV) include the introduction of new pedestrian signals, which traffic departments can use to replace the old type. The new alternative pedestrian signals consist of a flashing yellow light above a green one. The flashing yellow light means: "You may cross at your own risk." The crossing must be conflict free when the light is green.

Pussycats is a new system, characterized by technical improvements, better adapted to the behavior and needs of pedestrians, particularly those of vulnerable road users.

The pedestrian display has been moved to the near side of the crossing (the Maastricht position), facing the oncoming traffic. A mat detector replaces the push button, with infrared sensors detecting the presence of pedestrians on the crossing.

These technical improvements make it possible to show the pedestrian green light for short periods, to cancel unused calls, and to adjust the clearance time for slow pedestrians and large groups. Because of the new position of the display, pedestrians cannot see the pedestrian signals while crossing. This could encourage the watching of possible oncoming traffic and could also prevent pedestrians from becoming concerned or worried about lights turning red when they are halfway over the crossing.

More than 1,000 pedestrians were observed. Their crossing and watching behavior was noted in relation to the different phases, traffic flows, and the presence of other pedestrians. Two hundred users of the crossing were interviewed to obtain more information on their understanding of pussycats. They were asked to compare the old crossing with the new one, in terms of safety and convenience.

Figure 5. *Mat detector.*   Figure 6. *Infrared detector.*

Conclusions
1.   *Operations and efficiency*: The installation of the mat detector revealed serious problems closely related to the condition of the soil in the west and north of the Netherlands. Peaty soils make installation as prescribed almost impossible. (In more recent studies, however, this problem has been countered by replacing the mat detector by an infrared detector. The detection of waiting pedestrians with an infrared detector proved to be very successful.)

2.   *Safety*: The number of crossers during blinking yellow is considerable, but not exceptional by Dutch standards. Forty-six percent of arrivers on blinking yellow also cross on blinking yellow. Only one aspect of Pussycats could make a difference. Pussycats is characterized by a very short green phase (only 7 seconds). A longer green phase could lead to more arrivers on green and green crossers.

Another aspect of the system, the "wait" lamp, which is not exclusive to pussycats, is also important. The chance of crossing on green increases when people arrive with the "wait" lamp on. As could be expected, increasing the necessary waiting time is related to more red crossing. Contrary to expectations, no relationship was found between the number of vehicles and red crossing.

Watching, as demonstrated by head movements, is considerable, particularly before crossing. Red crossers are more careful. The Pussycats position of the display, on the side of oncoming traffic, seems to increase watching in the direction of oncoming traffic.

Most people questioned (87%) said they felt safe while crossing, but Pussycats was not responsible for this. The old system was not found to be safer then Pussycats. Reasons for unsafe feelings were sometimes related to Pussycats, such as the position of the light on the near side. It is suggested that information about the operation of the infrared detectors could prevent unsafe feelings relating to the pedestrian display. The most important safety advantage for vulnerable road users is the adaption to slow pedestrians.

3.   *Convenience*: Answers on the function of the mat show insufficient understanding, but the video survey shows that people know how to get a green signal if they intend to cross on green. There are no indications that the short green period (7 seconds) bothers the pedestrians. This might be expected, as the audible signal provides an efficient warning. The clearance time period extended by the infrared detectors proved to be at least 3 seconds too short, but hardly any complaints were made about this by the interviewed pedestrians.

The position of the pedestrian display at the near side of the crossing is regarded as a negative point. Two factors could improve the situation. First, if people know that an infra-red detector protects them from passing traffic, the unpleasant feelings linked to not seeing the display turn red could be tempered. Secondly, many people say that they are not use to such a position. Longer experience, covering more sites, could alter the situation.

2.6. **Carsten et al. (1992)**

This report summarizes work undertaken as part of the 3-year European Community DRIVE program that began early in 1989. The aim of the project was to examine the feasibility of developing a traffic system that meets the need of vulnerable road users (pedestrians and cyclists) both in terms of travel and safety. There are indications at present that the development of advanced traffic systems such as those envisaged by the DRIVE program as a whole, may have detrimental effects on pedestrians and cyclists. Most current developments are exclusively directed at the improvement of the safety and efficiency of motorized traffic and tend to neglect the position of vulnerable road users (VRU's). As a result, such systems may have negative safety and mobility effects for vulnerable road users that can seriously impair the positive effects on the traffic system as a whole.

Given the nature of the participants in the project, three countries were chosen as the basis for the work, namely the United Kingdom, Netherlands, and Sweden. The initial stage of the work was to examine the problems faced by vulnerable road users in these countries, and also in one urban area within each of the countries, where it is intended that the modeling and experimental work described above would be based. These urban areas were Bradford in the United Kingdom, Groningen in the Netherlands, and Växjö in Sweden.

One of the two principal tasks of the project was to prepare a traffic model, incorporating vulnerable road users. In addition to performing network assignment, the project would attempt to translate the information on flows of various classes of road users provided by the model into prediction of conflicts and hence provide some indication of safety effects.

The model requires as inputs real-world data on motor vehicle, cyclist, and pedestrian flows as well as on the route choice criteria for the various modes. To calibrate the model, data on behavioral response to modifications in the network are fundamental.

To achieve a wide range of behaviors and environments, data collection were carried out in each of the three countries (the Netherlands, Sweden, and the UK). Besides the collection of data, two main types of experiments were carried out. Firstly, two experiments studied the microwave detection of pedestrians (UK, Sweden). Secondly, an observational study was undertaken at an intersection in Groningen (the Netherlands) to test the potential of a system that gives car drivers prior warning when a cyclist approaches an intersection on a parallel bicycle path.

The DRIVE project has achieved three major pieces of work:

1. It has carried out extensive studies of vulnerable road user behavior in real-world situations to establish the factors underlying route choice and crossing strategies (where to cross). It has also begun the work required to establish the factors underlying crossing behavior (when to cross).
2. It has carried out diverse experiments using RTI (Road Transport Informatics) detection devices to alter the interaction of vulnerable road users with motorized traffic. Most of these have used the detection devices to alter signal timing in ways that are more responsive to vulnerable road users presence, but the project has also examined the potential for using the detection devices to activate warning signals that alert the driver to the presence of vulnerable road users.
3. It has developed a set of simulation models that represent, albeit in summary form, the movement of pedestrians around a street network and the safety consequences of the various road crossing flows. Another set of simulation models have been built to represent pedal cyclist behavior at junctions.

3. Traffic calming for pedestrians

The recent stagnation in further reduction of road accidents, insufficient results of existing policies to improve road safety, and the rather curative nature of these policies induced the wish to renew and improve road safety policy in the Netherlands. This new approach is called a sustainably safe road transport system.

This system has an infrastructure that is adapted to the limitations of human capacity through proper road design, vehicles fitted with ways to simplify the task of man and constructed to protect the vulnerable human being as effectively as possible, and a road user who is adequately educated, informed and, where necessary, controlled. As to the infrastructure, the key to arrive at sustainable safety lies in the systematic and consistent application of three safety principles:

- Functional use of the road network;
- Homogeneous traffic streams; and
- Predictability for road users.

Applying all three principles does have a preventative character to preclude as much as possible the incidence of accidents. A functional use of the road network primarily calls for establishing the intended function of every road. The present multifunctionality of roads leads to contradictory design requirements. Therefore, in a sustainably safe infrastructure, every road is appointed only one specific function. Pure through roads, pure distributor roads, and pure access roads.

In this chapter four studies are reviewed in which the effects of infrastructural measures (with an emphasis on traffic calming) on pedestrian safety are described. In the first study (Slop & Van Minnen, 1994; see Section 4.1), a comparison is made between a sustainably safe layout from the perspective of motorized traffic and a sustainably safe layout from the perspective of pedestrian and cycle traffic. The three other studies (see Section 4.2 - Section 4.4) describe analyses of traffic accidents with pedestrians before and after the construction of infrastructural traffic calming measures.

3.1. Slop & Van Minnen (1994)

Up to now, the concept of sustainable road safety was mainly elaborated from the perspective of motorized traffic. Policy aims such as more concern for vulnerable road users and promoting bicycle use, calls for proportional attention to pedestrian and bicycle traffic. To that end, this report sets forth the principles of sustainable safety, elaborated from the perspective of these two categories of road users.

Special attention is paid to the matter of incompatibilities between the perspective of motorized traffic and that of vulnerable road users. Subsequently, the general considerations are concretised by implementing them, on paper, in a trial area located in the centre of Gouda (the Netherlands).

Comparing the elaborations from the perspective of motorized traffic, bicycle traffic, and pedestrians, it comes out that the plans show great correspondence. The monofunctional road categorization for motorized traffic leads to solutions that also proved to be favorable for pedestrians and cyclists:

- Reducing the amount of motorized traffic on main roads;
- Separating traffic modes on main roads;
- Reducing the amount of motorized traffic in city centers, and providing parking space on the outskirts of the city centers;
- Replacing controlled intersections by roundabouts;
- Providing tunnels and bridges for cyclists and pedestrians to cross main roads.

Only in a few separate cases, the needs of motorized traffic and pedestrian/bicycle traffic can lead to conflicts of interest. These conflicts usually don't result in negative effects on traffic safety. The observed correspondence could very well mean that an elaboration in which all traffic modes are taken into consideration, will produce good results.

3.2. Dijkstra & Bos (1997)

This report is the Dutch contribution to the study carried out in several European countries as organized by the European Automobile Manufacturers Association. The report presents accident data on 173 sites in several Dutch cities before and after small-scale measures were introduced. The measures concern several types of pedestrian street crossing facilities and 30 km/h area implementations. Emphasis is given pedestrian safety effects. In the analysis a distinction is made between location measures (43 sites) and area measures (130 sites). Measures studied were:

Location measures
- narrowing.
- narrowing / small bicycle paths.
- narrowing / pedestrian waiting strips.
- median island.
- median island / lanes bending outwards.
- median island / axis realignment.
- median island / double axis realignment.
- median island / bus stop.
- junction size reduction.
- junction median island.
- roundabout.

Area measures
- 30 km/h signs only.
- road humps only.
- road humps / narrowings.
- road humps / axis realignments.
- road humps / other measures.
- road humps / narrowings/ axis realignments.
- road humps / narrowings/ other measures.

- road humps / axis realignments/ other measures.
- road humps / narrowings/ axis realignments/ other measures.
- road humps / street closures/ narrowings or axis realignments.
- narrowings or other measures (without humps).
- axis realignments / narrowings or other measures (without humps).

Conclusions about location measures
With regard to the number of all injury accidents, it can be observed that apparently about 50 percent of the location measures has contributed positively to traffic safety, whereas the other 50 percent has had a negative safety effect. Only the junction measures (junction size reduction, junction median island, roundabout) seem to consistently generate less accidents. The larger effects, however, mostly are based upon few data and therefore are not very reliable.

The overall result of the measures is slightly positive for traffic safety. With respect to pedestrian safety the situation is worse. Except in case of a roundabout, the numbers of both pedestrian involved accidents and pedestrian victims have increased after the measures, albeit effect estimates are rather uncertain because of generally small data bases. The overall result of the measures is anyhow negative for pedestrian safety.

Conclusions about area measures
Accident data show that all measures were coupled with a diminished number of all injury accidents. In one case, however, no effects can be determined. No comparison could be made because there were zero accidents in the before period because of the short study period. It is striking that even the simple use of 30-km/h (19 mi/h) signs only seems to have a considerable positive effect on general safety. The total number of accidents decreased after the introduction of the 30-km/h (19 mi/h) signs, but the number of pedestrian accidents increased.

Half of the measures had a positive effect on pedestrian safety. In the other half of the cases, pedestrian safety became worse. Nevertheless, the overall safety effect of the area measures is positive for pedestrians, because the positive effects (decrease of accidents) proved to be larger than the negative effects (increase of accidents).

The authors indicate that small numbers and the consequent lack of reliability of the effect estimates were a main problem in this research. Therefore, valid conclusions can at most be drawn at a more overall and general level. In fact, following a more strictly statistical approach, it is obvious that but a very few results may possibly be tested significant at a level of better than 90 percent. Within this context, it is noticed that area type measures seem much more effective than location type measures. This is true with respect to all injury accidents as well as pedestrian involved accidents. Also it is true regarding the number of pedestrian victims and the severity of pedestrian injuries. Furthermore, it is found that area wide measures are more safety effective if taken at sites with larger volumes of street crossing pedestrians.

3.3. Vis & Kaal (1993)

The 30-km/h (18 mi/h) zones are supposed to improve road safety and quality of living in areas which predominantly serve a residential function. During a previous study of 15 experimental 30-km/h zones, it was concluded that the total number of accidents after introduction of the measure had dropped by 10 to 15 percent. With respect to the number of injury accidents, there were indications that the reduction may have amounted to double that figure. Because of the limited scale of the study however the effects demonstrated a large spread.

Figure 7. *30 km/h zone.*

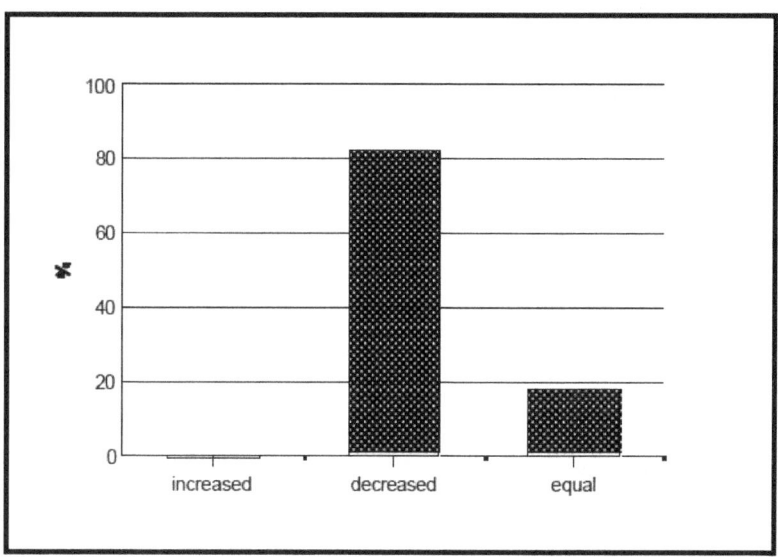

Figure 8. *Changes in speed after reconstruction as 30-km/h zone.*

In this follow-up study, the effect on the number of injury accidents in a large number of 30-km/h zones was more specifically determined. In this study no special attention is paid to the traffic safety of pedestrians. In earlier studies however, it was concluded that most injury accidents in residential areas concern accidents in which pedestrians and cyclists are involved. A decrease in accidents in residential areas therefore most likely leads to a decrease in the number of injured pedestrians and cyclists.

Of 151 30-km/h zones, 660 injury accidents were recorded: 417 before introduction of the measure and 243 during the follow-up period. To enable correction of effects which were not associated with the measure studied, all injury accidents inside the built-up area were collected for the same municipalities over similar periods (control areas).

Following correction based on the trend shown in the control areas, it was determined that the number of injury accidents in the 30-km/h zones had dropped by 22 percent (±13%).

Again, the effect on the number of injury accidents still demonstrated a large variation. Taking into consideration the (average) results, however, the measure can certainly be considered successful. Over half of the surveyed municipalities had not yet commenced work to realize 30-km/h zones even though the survey held among officials from the traffic departments of the municipalities in question demonstrated that a positive attitude prevailed. Intensive stimulation to foster implementation of 30-km/h zones on a broader scale is therefore recommended, while further study into the causes of the reticence shown by many municipalities would be useful.

Furthermore, it is advisable to check if the quality of the applied countermeasures in the 30-km/h zone are functioning as planned and if this is not the case, to find out why, to avoid this in the future. It has been shown that those areas that are designed as 30-km/h zones tend to carry a lower volume of motorized (through) traffic.

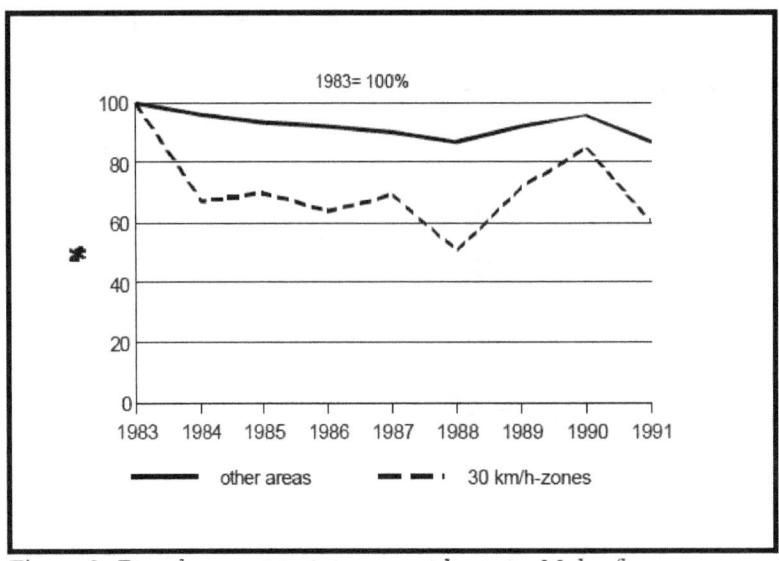

Figure 9. *Development in injury accidents in 30-km/h zones.*

3.4 Kraay & Dijkstra (1989)

Analysis of registered accidents in the Netherlands proves that traffic accidents inside built-up areas mainly is a problem of cyclists and pedestrians conflicting with motorized traffic. Further, an important observation is that only 20 percent of the accidents actually take place on residential streets and 80 percent on the main roads. From the point of view of traffic safety, the greatest results in improving safety can therefore be expected from measures on the main roads. Within residential areas traffic accidents are mostly not concentrated on black-spots but take place scattered over the entire area.

To improve traffic safety in residential areas in most cases, technical measures are needed to influence traffic behavior in a positive way. Through traffic must be kept out of residential areas as much as possible. Motorized traffic having its origin or destination within the residential area must adapt its behavior to the residential character. This implies that driving speeds may not exceed 20 to 30 km/h.

An area wide approach is, regarding the nature of the problems, far more to be preferred than improving several separate locations. Experiences in the past have shown that a strict differentiation of roads and streets according to their function in the network is a good way of improving traffic safety in residential areas. Analysis of traffic accidents in redesigned residential areas, as carried out in this study, proves that structural redesign has a positive effect on traffic safety. In residential streets in these redesigned areas, the amount of injury accidents per vehicle-km has decreased approximately 70 percent. On main roads and arteries in these areas, a decrease of approximately 20 percent was found.

4.Children

This chapter discusses a variety of different types of research. Two studies describe the mobility and freedom of movement of children in relation to traffic safety. In the first (Van der Spek & Noyon, 1993), the authors try to explain the decrease in the number of accidents with children by the supposed decrease in the freedom of movement. The second study (Dutch Pedestrians Association, 1993) consists of school surveys on traffic safety in school zones and in school routes. Results show that the freedom of movement of the children has decreased over the last years. This decrease is explained by the negative judgment of traffic safety on school routes both parents and teachers give. Parents no longer let their children go to school independently, but bring their children themselves.

The study *Driving Strategies Among Younger and Older Drivers When Encountering Children* (Lourens et al, 1986) describes observations of the behavior of car drivers (and deficiencies in that behavior) in traffic situations in which children are involved. A study by J. Brinks (1990) describes the behavior of children in traffic and the deficiencies in that behavior.

The study *Pedestrian Injury Prevention* (Molen & Linden, 1987) describes the development of a traffic training program for children called "Crossing the Street." Next, Douma (1988) evaluates the training program "Crossing the Street."

4.1. Van der Spek & Noyon (1993)

Over the last 20 years, the total number of cars in the Netherlands increased by 85 percent. A growth from 3 million cars in 1972 to 5.6 million in 1992. On the other hand, the total number of fatal and injury accidents in traffic decreased over the same period. The total number of fatal accidents decreased by 60 percent and the total number of injury accidents decreased by 30 percent.

For traffic safety the year 1972 is a turning point. Until that year the increase of the number of victims caused by traffic accidents was equal to the growth in traffic volumes. Since 1972 however, the number of victims is decreasing almost every year, whereas the yearly increase in traffic volumes is still going on. This beautiful result could be caused by technical measures taken over the last 20 years. The life threatening influence of the car tamed by technical measures. The opposite, however, could also be true. Man has adapted himself to the negative influence of the car. Man tamed by the car.

This study tries to explain this paradox; not only by examining the effects of technical measures but by examining other factors. Examined is whether the freedom of 4 to 12 year old children has decreased, and to what extent this decrease in freedom of movement can be seen as an explanation of the decrease in the number of traffic injuries.

One generation ago children played outside more often than nowadays. Children were outside their homes most of their time and had more freedom of movement. The games children played by that time, demanded a lot of (public) space. Nowadays, playing outside is not obvious for a lot of children: 12.6 percent of the children questioned, almost never played outside, and nearly 30 percent of the children played outside no more than three times a week.

Playing in the streets is hardly the case anymore. Children nowadays mostly play outside in backyards and squares. In many cases children only play outside under supervision. When children are playing with friends, 44 percent of them are brought and taken home by the parents. Therefore their life is more and more organized and their freedom of movement restricted. If children go to clubs or sporting clubs, 65 percent of them are transported by the parents.

Most children are allowed to play outside only near the house from the age of 5 or 6 years. Moving further away from the house is only allowed from the age of 8.

The relation between the freedom of movement of children and the opinion their parents have about traffic safety in the neighborhood proves to be very clear. If parents give a positive judgment of traffic safety, 55 percent of the children are allowed to go to school unsupervised. When the parents give a negative judgment of traffic safety in the neighbourhood, only 22 percent of the children are allowed to go to school unsupervised. When technical measures are taken to improve traffic safety in the neighbourhood, the judgment of traffic safety by the parents improves, but remains insufficient.

The decrease in traffic accidents with children, while traffic volumes increase, is not very strange. Children don't play outside as much as they used to, mainly because of unsafe traffic. When children go outside to go to school or to play with a friend, they usually are supervised by parents. Traffic hasn't adjusted to the children, but children have adjusted themselves to traffic.

Figure 10. *Reconstructed residential area.*

4.2. **Dutch Pedestrians Association (1993)**

For several years, the Dutch Pedestrians Association has conducted a school survey (kindergarten and primary school) on traffic safety in school zones and school routes. Here the results of the last survey in 1993 are presented.

Developments in traffic are unfavorable for the traffic safety of school children. The volumes of motorized traffic have increased by 10 percent over the last 5 years. Several schools noticed a deterioration of the behavior of traffic participants. This increasingly places higher demands on the children in walking to and from school.

Fewer children come to school independently. In 1970, the average age on which as many as 80 percent of the children came to school independently was 6 years. In 1993 that age was 8 years. The average distance children had to travel to and from school has increased, leading to an increase in hazardous situations in the school route. Because more and more children are being brought to school by car, traffic volumes increase and from that unsafety increases. This unsafety causes other people to bring their children to school by car too.

School routes are unsafe. Fourty-one percent of the children have to pass busy roads and unsafe locations on the way to school. There seems to be no improvement of the number of traffic accidents on school routes.

Measures to improve the safety of routes to school remain necessary. At more than 50 percent of the schools, measures to improve traffic safety have been undertaken. At more than 30 percent of the schools, measures have not yet been taken, though they should have been. Twenty-eight percent of the schools claim that undertaken measures proved to be unfavorable for traffic safety.

Parents, schools, and government have a shared responsibility. Parents and schools can locate problems, parents can monitor their own behavior and governments have to take care of safe routes.

4.3. Lourens, Van der Molen & Oude Egberink (1986)

This report presents the results of a study into how drivers *say* they behave, and how they actually behave in traffic situations in which children are involved. An analysis was made of the most important types of encounters in which drivers become involved in accidents with walking, playing, or cycling children.
On the basis of accident surveys and psychological theories on information processing, it was assessed by means of a questionnaire concerning their own behavior in these situations, as well as their expectations about typical child behavior. Actual behavior of drivers in these situations was investigated by assessing video recordings of their behavior in driving a 1-hour standard track through residential areas.

The most important findings of the study are:
- Younger drivers report their own risky driving behavior more often than older drivers;
- Older drivers underestimate their own speed more often than younger drivers;
- Female drivers underestimate their own speed more often than male drivers;
- Young female drivers proved to score less on the driving task than other drivers;
- There proved to be no relation between driving experience and results in the tests;
- While thinking out loud during the driving test, older drivers prove to give more evaluative judgments, while younger drivers prove to give more detection judgments;
- While thinking out loud during the driving test experienced, drivers give more decision judgments than less experienced drivers.

The relatively poor score of female drivers in this study could be caused by a coincidental non-representative construction of the tested group. The authors find it premature to connect any conclusions to these findings. In the report implications for the contents of mass media campaigns and their evaluation are discussed.

4.4. Brinks (1990)

This study mainly focuses on traffic safety of young cyclists. The results of the study however also apply to the skills of young pedestrians.

In the Netherlands, many cyclists of 12 to 16 years of age are involved in accidents or near accidents. Various studies of basic cycling skills and functional abilities required for safe cycling behavior indicate that these skills and abilities are for the most part adequately mastered. So, other factors that contribute to (un-)safe cycling behavior must explain the high accident involvement. From a cognitive point of view, the knowledge of traffic rules and signs, the knowledge of (normative) rules governing complex maneuvers and also processing environmental information and linking this information to the proper actions are presumed to contribute to accident involvement. Moreover, attitudinal and motivational issues (including risk acceptance) are pointed out as important factors in accident involvement, particularly in the age group concerned. Our understanding of the way these factors link to accident involvement is increasing. However, little is known about to what extent these factors are mastered in the age group concerned.

In the framework of an evaluation research project concerning the implementation of traffic educational materials, the author extensively investigated the initial situation of 12 to 16 year old children with regard to most of the factors mentioned above.

The investigation shows that there are severe deficiencies with regard to the knowledge of priority rules, particularly when right of way is not indicated by signs or road marks. Also the knowledge of (normative) rules governing complex maneuvers (such as turning left at an intersection) is inadequate. The same goes for anticipating risks and reacting to these anticipated risks in a safe manner.

With regard to attitudinal issues, it is found that attitudes towards safe traffic behavior are cause for concern. It seems that for 12 to 16 year old children violations of quite dissimilar nature form a sort of conglomerate. Adults (i.e., teachers) on the other hand, appear to differentiate their attitudes with regard to violations in specific situations. This might mean that whereas adults judge their actions on an occasion by occasion bases guided by expert knowledge, 12 to 16 year old children still lack this cognitive skill. The consequences that the findings of this study may have for traffic educational objectives and programmes are discussed.

4.5. Van der Molen & Van der Linden (1987)

Two major types of measures are dealt with for countering pedestrian injury in residential areas: child pedestrian training and the construction of (residential) yards (in Dutch: 'woonerven').

At the request of the Dutch Ministry of Transport, a child pedestrian training program for 4- to 6-year-old children has been developed at the Traffic Research Centre of the University of Groningen. The major aim of the traffic training program, is to ensure that the children will cross more safely in the streets where they generally play or walk to kindergarten. It is not, however, the aim of the program to encourage parents to let their children cross on their own more frequently. All children in this age group do cross some roads in their neighbourhood on their own, however. These are generally very quiet roads without zebra crossing or traffic lights. It is in these very quiet streets however, that the children in this age group become involved in accidents.

On the basis of research, the authors concluded that for this age group (4- to 6-year-olds) the following road-crossing tasks are the most important:
    a.   Crossing at midblock without visual obstacles;
    b.   Crossing at midblock from between parked cars;
    c.   Crossing at intersections (without visual obstacles).

Task (c) is more important for the 5-year-olds, as younger children cross less frequently at intersections. Moreover it is a relatively difficult task for the younger children.

The three road-crossing tasks include a list of actions and decisions that have to be made for a safe completion of the task. These actions and decisions are used as training objectives in the traffic training program, as described in Table 4 on the following page.

The actual training of the children in the street is done by the parents. Through the school, they receive an instruction booklet in which they can read in great detail how to carry out the training in each of the three tasks. In a film at a parents-meeting at school the following training steps are demonstrated:

1. Modeling: the parent demonstrates the desired behavior to the child.
2. Practice together: parent and child practice the desired behavior together.
3. Practice alone: the child tries to carry out the desired behavior under supervision of the parent.
4. Observation and reward: the parent observes the child and rewards the child for each behavioral element performed correctly.

The parents should carry out training in each task for about 15 minutes a day for 1 week. When the children were tested by test assistants after the training period, most of them were able to perform almost all behavioral objectives correctly. When tested half a year later this was still the case. It can therefore be concluded that the program is very successful in establishing the desired behavioral repertoire.

a. Crossing at midblock without visual obstacles	b. Crossing at midblock from between parked cars	c. Crossing at intersections (without visual obstacles)
walk to the curb at normal speed	walk to the curb at normal speed	walk to the curb at normal speed
stop before the curb	stop before the curb	stop before the curb
	at the curb look inside the parked cars	
	stop at the line of vision	
	stand near the right-hand car	
look left at the curb	look left at the line of vision	look left at the curb
		look ahead at the curb
look right at the curb	look right at the line of vision	look right at the curb
		look behind at the curb
wait if traffic approaches	wait if traffic approaches	wait if traffic approaches
start to look out again when traffic has gone	start to look out again when traffic has gone	start to look out again when traffic has gone
cross at normal speed and right angles	cross at normal speed and right angles	cross at normal speed and right angles

Table 4. *Training objectives for three pedestrian tasks.*

The children were also observed unobtrusively before and after the training period, while playing outside with their friends or walking to kindergarten unsupervised. The improvements in performance were significant. On the other hand however, the observational data show that under normal conditions children do not behave according to their newly acquired abilities.

4.6. Douma (1988)

Douma (1988) discusses the results of a study into the effects of the introduction of a traffic education program for young children. In a pilot study, an interview method was developed to get a better insight into the way in which young children were involved in accidents. The amount of accidents with children of the test group proved (luckily) too small, and registration of the accidents too incomplete to use in the evaluation. The evaluation of the education program therefore is being made by using the results of brief interviews of the parents. The results show that the education program called "Crossing the road" was effective.

In the study some groups of children proved to be more accident prone than others. The chance of getting involved was found to depend on a set of personal characteristics, backgrounds, and exposure. Boys, children of foreign parents, and children that were allowed to play longer outside proved to be more accident prone. Traffic safety measures should be focussed on the accident prone groups of children.

5. Elderly traffic participants

Both studies reviewed in this chapter describe the problems of elderly traffic participants by analyzing traffic accident data and mobility data. The defined problems are then explained by cognitive and physiofunctional changes bound up with aging. Possible measures to reduce the problems are described.

5.1. Van Wolffelaar (1988)

Van Wolffelaar (1988) reviews the problems of elderly traffic participants as derived from statistical, experimental, and gerontological publications. The presented data include changes in mobility, accident involvement, and behavioral problems of car drivers, bicyclists, and pedestrians. Cognitive and physical changes as a consequence of aging are reviewed, and finally some educational objectives for elderly traffic participants are derived from theoretical possibilities of behavioral improvements.

Statistical data indicate that there is a general decrease in mobility among elderly. This applies especially for the distances driven by car, mainly as a consequence of decreased professional activities.

Per distance traveled, however, elderly people are increasingly involved in traffic accidents. Particularly the proportion of victims among bicyclists and pedestrians increases dramatically with age, mainly because of the greater vulnerability because of their unprotected traffic environment.

A striking increase of accident rate is observed in conditions of high traffic complexity and time pressure. This finding is in conformity with results form gerontological studies concerning cognitive and physical functional deteriorations among elderly. These indicate an age-related decrease in functional capacities from which increased problems may be anticipated in complex traffic situations, demanding fast and accurate perceptions, decisions, and responses.

First of all, there is a deterioration of sensual perceptions (vision and hearing). Furthermore, the most noticeable characteristic of elderly traffic participants proves to be a slowing down of behavioral performance in both motorial functions (muscles and joints) and psychological functions.

The main issue for education of elderly traffic participants, therefore, should be learning to cope adequately with the effects of aging. The main targets of traffic education of elderly traffic participants should be:

1. Primarily education: Improvement of the knowledge of traffic rules and traffic skills;
2. Secondary education: Improvement of the knowledge of the effects of aging, learning to cope with loss of function (compensation), and acknowledgement of the need of a good mental and physical condition.

5.2. Wouters (1991)

Wouters (1991) provides an overview of recent data on the road safety of elderly people in the Netherlands. Compared with the 30- to 50-year-olds, the paper shows that:

- The road hazard magnitude of elderly road users is higher;
- Elderly people have more serious accidents;
- Elderly people, particularly as pedestrians and cyclists, have a considerably high risk of injury accidents.

The road hazard of elderly people is mainly caused by three interrelated factors. These factors include:

1. Physical vulnerability.
2. The loss of mental and physical function. With growing of age perceptive, cognitive. and motor skills decrease. In traffic this can lead to poor vision in dark and twilight, decreasing of the ability to estimate speed and distance, and decrease of hearing. Complex situations in traffic can cause problems in selection of information and decision making. Different decisions can no longer be taken virtually simultaneously, but only successively.

3. A mobility decrease. The decrease in mobility is strongly related to socio-economic factors such as decreasing family-size and retirement. Other factors can be the fear of not being able to come along in traffic, fear of own vulnerability in traffic, or feeling unsafe in traffic.

The decrease in mobility leads to further deterioration of mental and physical skills and loss of routine. As a result of this, participation in traffic becomes more and more dangerous.

The possibilities for breaking through this vicious circle are as follows:

- Slowing down the loss of mental and physical function of elderly people by either maintaining or improving their traffic skills. This means that they should be stimulated to keep on participating in traffic.
- Other road users should have more consideration for both the possibilities and limitations of elderly people.
- The traffic situations should be modified in such a way that elderly people can participate in traffic in both a satisfactory and safe way.

6. Provisions for disabled pedestrians

The two publications reviewed in this chapter both describe measures and provisions for disabled persons. Neither of the publications describe research on those measures. The first publication is a manual for infrastructural measures for safe and independent traffic participation by disabled persons. The second publication describes a device with which pedestrians can double the duration of green light for pedestrians at signalized crossings. This device is used by elderly and disabled pedestrians in the municipality of Enschede.

6.1. Prikken & Gerretsen (1988)

One of the aims of the policy of the Ministry of Traffic and Transport is to improve the provisions for safe and independent traffic participation by disabled people. To make an inventory of complaints and existing problems, a written interview was held among a selection of organizations of handicapped people (Prikken & Gerretsen, 1988).

Complaints of handicapped people mainly concern problems experienced in city centers and shopping centers. Problems that handicapped people encounter can be divided into the following groups:
- Route difficult to traverse;
- Problems reaching certain destinations;
- Accessibility of destinations;
- Usability of provisions or destinations.

The necessary bottlenecks which should be inventoried are road sections and streets, crossing places, junctions, roundabouts, squares, shopping areas, and traffic restrained residential areas. The manual pays attention to organisational aspects, but also gives technical solutions for problem situations. Traffic safety effects of the given measures are not studied or reported.

Research shows that in the bigger cities, structural arrangements are made by the organizations of disabled people. In smaller cities problems are mostly solved by ad-hoc solutions. The problems are then mostly under-estimated and applications are mostly not executed well.

6.2. Municipality of Enschede (1992)

In its report, the Municipality of Enschede describes the results of an evaluation of an experiment in the municipality of Enschede with so-called pedestrian transmitters. The pedestrian transmitter is a device by which pedestrians can double the duration of green for them at signalized crossings. It also activates a sound signal on the traffic lights, indicating red and green for pedestrians. Moreover, the pedestrian transmitter stops all

Figure 11. *Guidance strip for visually handicapped.*

other directions when the pedestrian light turns green. All directions turn red on green for pedestrians. The pedestrian transmitter can be used by people who need more time to cross than the average pedestrian (elderly and handicapped).

In the experiment in Enschede, 13 signalized intersections were adapted to the use of pedestrian transmitters. This number is being extended after successful completion of the experiment.

In the evaluation, users of the pedestrian transmitter and contact persons of homes for the elderly were interviewed. Results show that more than 50 percent of the users use the pedestrian transmitter several times a week. Nearly 70 percent of the users claim that they would not take the same route if they didn't possess the pedestrian transmitter. In general, the users of the pedestrian transmitter are very pleased with both the effect and the functioning of the pedestrian transmitter.

7.Car front-impact requirements

The three studies described in this chapter discuss two different aspects of a car's front impact requirements. The first study (Van Kampen, 1994) concerns a comparison of both costs and benefits of the implementation of car's front-impact requirements in the Netherlands. The benefits have been calculated by determining the value of average costs of killed and injured victims, combined with the estimated casualties spared by a car's front-impact requirements.

The other two publications relate to the same study (parts I and II). The publications describe the development of test methods for evaluating pedestrian protection for passenger cars. The main result of the study is the development and calibration of computer models to describe the severity of injuries of pedestrians being hit by a passenger car, under different circumstances and with different car designs.

7.1. Van Kampen (1994)

In the Netherlands the traffic safety of pedestrians and cyclists has been a major concern for many years, though both the annual number of pedestrian casualties and cyclist casualties have decreased during the past 10 to 20 years, as in almost all European countries.

Dutch policy aims at further reducing these numbers. The proposed measure, introducing tests regarding the front end of cars, is strongly supported by the Dutch Ministry of Transport, since it is expected that both pedestrians and cyclists will benefit.

To establish a stronger (international) base for this purpose, the Dutch Ministry of Transport has agreed to have SWOV carry out a cost/benefit study on the subject, of which the general design should be comparable to similar studies, already carried out by TRRL (UK) and BASt (Germany) to compare results. In the report by Van Kampen (1994), the Dutch cost benefit analysis carried out by SWOV is described.

The scope of the problem is derived from Dutch national accident data. The annual number of casualties, relevant to the problem of collisions with car front-ends, is at least 6,500 (pedestrians and cyclists). Nearly 200 of these casualties were killed, while 1,900 were hospitalized. It is certain that the remaining number of other injured (slightly injured) is in reality far greater than the 4,400 registered casualties, because of the problem of under-registration.

In another part of the study, gross costs pertaining to casualties have been calculated. This resulted in a 1991 value of average costs per fatality of about 900,000 guilders (415,000 ECU's); the costs per hospitalized are about 115,200 guilders (53,000 ECU's); costs per slightly injured are 28,800 guilders (13,300 ECU's).

The expected effectiveness of the proposed measure has been derived from in-depth accident data, following the model used in the BASt-study, mentioned before. Using this effectiveness data, as well as the cost data and the national accident figures, Dutch benefits of the proposed measure have been calculated, their total number being more than 750 casualties spared (of whom 11 fatalities, 263 hospitalized). In 1991 money value, these annual benefits amount to 24,800,000 ECU's.

These benefits are the result of the compliance of new cars to the proposed measure. Assuming that each year some 500,000 new cars complying to the measure replace the same number of older cars, the cost per new car may be up to 50 ECU's to keep a positive cost/benefit ratio. In view of extra cost expectations for new cars, complying to the measure as reported in the TRRL-study mentioned above, this means that a positive ratio of benefits over costs of 3:1 is feasible. It is concluded that implementation of the proposed measure will be of great benefit for the Netherlands.

7.2. Janssen & Nieboer (1990) / Janssen, Goudswaard, Versmissen & Van Kampen (1990)

The European Experimental Vehicles Committee has set up a Working Group to assess and develop test methods for evaluating pedestrian protection for passenger cars. The methods are subsystems tests to the bumper, the bonnet leading edge, and the bonnet top.

Test conditions appropriate for vehicle-to-pedestrian impacts of up to 40 km/h are considered, with adjustments made to allow for the influence of the vehicles frontal shape.

Computer simulations using the MADYMO CVS program are performed by TNO to gain a better understanding of the complex kinematics of a pedestrian accident. The influence of vehicle shape and pedestrian anthropometry is analyzed, as well as the influence of vehicle speed, vehicle stiffness, and walking position of the pedestrian.

From the 45 basic simulations and 18 additional simulations, it was shown that some vehicle parameters considerably influence the pedestrian responses, while some parameters hardly influence the responses. Furthermore, it was shown that the responses of the 5th percentile female are within the ranges of responses of the 50th percentile male and 6-year-old child. These simulations have shown that the selected protection criteria, for instance the bending moment in the upper and lower leg and the knee bending angle, are very well able to discriminate between different vehicle shapes and stiffnesses.

Based on these conclusions, test conditions are proposed for the subsystems tests on the bumper, bonnet leading edge, and bonnet top.

In most European countries and in the United States, the proportion of pedestrian fatalities caused by collisions with motor vehicles ranges from about 15 to 30 percent of the respective national traffic death toll. Given this fact, it is considered necessary that this problem be attacked by both pre-crash (accident prevention) and crash (injury prevention) strategies. This literature review is concerned with the latter strategy.

Both in the U.S. and in Europe (EC) long-term research strategies supported by governments have been followed by legislation in the near future. Practically speaking this means that safety requirements will be established concerning the (subsystem) testing of the front ends of cars with respect to collisions with pedestrians.

Ample proof exists that such pedestrian safety requirements are feasible for newly developed cars, while there is also proof that considerable benefits are possible through minor changes of current car design.

The European situation differs from the American with respect to accident and vehicle characteristics, but the offer of the United States to make use of their still growing experience, based on their ongoing pedestrian research program, should be accepted.

One of the grounds for manufacturers opposed to the proposed new requirements is their concern that these may conflict with existing requirements. Most of the examples of this conflict focus on bumper regulations. These existing requirements aim at the reduction of damage and damage repair cost in minor (low-speed) collisions.

Evidence has been found that it is indeed difficult to combine the two sets of regulations, especially when all detailed requirements of the current bumper regulations have to be met as well as future pedestrian requirement. Some compromise between the two sets of requirements could therefore be expected. Also ample evidence is found, however, indicating that application of new materials, both lightweight aluminium and various other kinds of energy absorbing material, will solve most of these problems, even without taking away the characteristics of individual car design. Further, evidence has been found indicating that such new designs will improve the outcome of car-to-car collisions, especially in cases of side impacts with regard to the overall damage.

The fears of car manufacturers that occupant safety may be impaired by the new requirements on pedestrian safety are at least theoretically unfounded. Mass differences between cars and pedestrians as well as the possibilities to make far better use of available crush distances in both current and new car design guarantee that occupant safety will not be impaired.

From the viewpoint of costs and effectiveness, the final assessments cannot be made since almost no real-world experience with these types of constructions exist and therefore could not be reported. However, effectiveness estimates based on experimental designs combined with known figures of the population at risk and the cost of medical treatment of injuries as well as the societal costs of fatalities, injured, and impaired, point to a positive balance between effectiveness and costs.

It is recommended, however, that both research institutes and car manufacturers stimulated by their respective governments combine forces and seek for still further improvement of car design. This is absolutely necessary in view of the still existing amount of crash incompatibility between motor vehicles of the same type and between different types of road users.

8. Summary

Analysis of reported accidents during recent decades showed that the number of pedestrians killed in a traffic accident in the Netherlands decreased in the eighties. This decrease came to an end in the nineties. The year 1996 however shows a remarkably positive development.

The stagnation in the decrease can hardly be explained by an increase in pedestrian mobility for this has remained relatively unchanged between 1980 and 1994 (5 to 5.5 billion km per year). In recent years the level of pedestrian mobility is somewhat higher.

Pedestrians mainly are killed in accidents with cars. Pedestrian safety therefore is also determined by the mobility of motorized traffic. This mobility of motorized traffic is still growing every year.

Children and elderly pedestrians prove to be the most vulnerable. Nearly 50 percent of the total number of killed pedestrians are older than 65 years. Their risk, expressed as the number of deaths per km, is also found to be very high: more than 100 deaths per billion km, compared to 27 on average for all age groups.

Next to the elderly, children 14 or younger are the second most vulnerable age group. The number of children killed in a traffic accident has however decreased more than in other age groups.

Pedestrian crossings and traffic calming measures

In the report, several studies on pedestrian crossings are discussed. Attention is paid to safety aspects of signalized and unsignalized pedestrian crossings, as well as to innovative measures meant to improve signalized pedestrian crossings.

Attention is also paid to studies in which the effect of infrastructure measures (with an emphasis on traffic calming) on pedestrian safety are described. A comparison is made between a sustainably safe layout from the perspective of motorized traffic and a sustainably safe layout from the perspective of pedestrian and cycle traffic. Other studies describe analyses of traffic accidents with pedestrians before and after the construction of infrastructure traffic calming measures.

Children and the elderly

A summary is given of two studies that describe the mobility and freedom of movement of children in relation to traffic safety. In the first study the authors try to explain the decrease in the number of accidents with children by the supposed decrease in the freedom of movement. The second study consists of school surveys on traffic safety in school zones and in school routes. Results show that the freedom of movement of children has decreased during recent years. This decrease is explained by the negative assessment of traffic safety on school routes both parents and teachers give. Parents no longer let their children go to school independently (which used to be very common in the Netherlands) but take their children themselves.

The problems of elderly traffic participants are described by analyzing traffic accident data and mobility data. The defined problems are then explained by cognitive and physiofunctional changes related to aging. Possible measures to reduce the problems are given.

Disabled persons

The two publications reviewed in chapter 7 both describe measures and provisions for disabled persons. Neither of the publications describe research on those measures. The first publication is a manual for infrastructural measures for safe and independent traffic participation by disabled persons.

The second describes a device with which pedestrians can double the duration of a green light for themselves at signalized crossings. This device is used by elderly and disabled pedestrians in the municipality of Enschede.

A car's front-impact requirements

The three studies described in chapter 8 discuss two different aspects of car front-impact requirements. The first study concerns a comparison of both costs and benefits of the implementation of car front-impact requirements in the Netherlands. The benefits have been calculated by determining the value of the average costs of killed and injured victims, combined with the estimated casualties spared by a car's front-impact requirements.

The other two publications relate to the same study (parts I and II). The publications describe the development of test methods for evaluating pedestrian protection for passenger cars. **THE MAIN RESULT OF THE STUDY IS THE DEVELOPMENT AND CALIBRATION OF COMPUTER MODELS TO DESCRIBE THE SEVERITY OF INJURIES OF PEDESTRIANS BEING HIT BY A PASSENGER CAR UNDER DIFFERENT CIRCUMSTANCES AND WITH DIFFERENT CAR DESIGNS.**

References

Brinks, J. (1990). *Traffic Related Knowledge, Attitudes and Risk Perception in Dutch Secondary School Children; Consequences for Traffic Education.* "Proceedings of Road Safety and Traffic Environment in Europe." VTI report 364, p. 31-43. Gothenburg, Sweden.

Boot, T.J.M. (1987). *Verkeersongevallen op de VOP en de GOP op kruispunten (Traffic Accidents on Signalized and Non-signalized Pedestrian Crossings).* SVT, Driebergen, Netherlands.

Carsten, O.M.J., Draskoczy, M., Ekman, L., Schagen, I.N.L.G. van, Sherborne, D.J., Tight, M.R. & Timms, P.M. (1992). *Drive Project V1031; An Intelligent Traffic System for Vulnerable Road Users.* Institute for Transport Studies, University of Leeds, United Kingdom.

CBS (Statistics Netherlands) (1992). *De mobiliteit van de Nederlandse bevolking (Developments in Mobility for Different Transport Modes in the Netherlands/Exposure to Risk in Traffic Participation).* Centraal Bureau voor de Statistiek CBS (Statistics Netherlands), Heerlen, Netherlands.

Dijkstra, A. & Bos, J.M.J. (1997). *ACEA — Dutch Contribution; Road Safety Effects of Small Infrastructural Measures with Emphasis on Pedestrians.* SWOV, Leidschendam, Netherlands. [Draft]

Douma, M. (1988). *Summatieve evaluatie van het verkeersoefenprogramma 'Oversteken.' (Summative Evaluation of the Traffic Evaluation Programme, "Crossing the Road.")* Rijksuniversiteit Groningen, Verkeerskundig Studiecentrum, Haren, Netherlands.

Gemeente Enschede (Municipality of Enschede) (1992). *Evaluatieonderzoek voetgangerszender (Evaluation of the Pedestrian Transmitter).* Enschede, Netherlands.

Janssen, E.G. & Nieboer, J.J. (1990). *Protection of Vulnerable Road Users in the Event of a Collision with a Passenger Car. Part I: Computer Simulations.* TNO Road-Vehicles Research Institute, Delft, Netherlands.

Janssen, E.G., Goudswaard, A.P., Versmissen, A.C.M. & Kampen, L.T.B. van (1990). *Protection of Vulnerable Road Users in the Event of a Collision with a Passenger Car. Part II: Sub-systems Test Method: Evaluation and Compatibility Study.* TNO Road-Vehicles Research Institute, Delft, Netherlands.

Janssen, W.H. & Horst, A.J. van der (1991). *Een evaluatie van 'knipperend geel' op geregelde voetgangersoversteekplaasten. (An Evaluation of "flashing yellow" at Signalized Pedestrian Crossings).* TNO-IZF, Soesterberg, Netherlands.

Kampen, L.T.B. van (1991). *Analyse van letselgegevens van fietsers en voetgangers. (Analysis of Injury Data of Cyclists and Pedestrians).* R-91-56. SWOV, Leidschendam, Netherlands.

Kampen, L.T.B. van (1994). *Cost-benefit Study Concerning Impact Requirements to Increase the Crash Safety of Pedestrians and Cyclists; Final Report.* R-94-31. SWOV, Leidschendam, Netherlands.

Kraay, J.H. & Dijkstra, A.(1989). *Veiliger woonwijken zijn mogelijk. (Safer Residential Areas are Possible).* In: Voor alle veiligheid; Bijdragen aan de bevordering van de verkeersveiligheid. The Hague, Netherlands.

Lange, M. de (1996). *De positie van de voetganger op oversteeklocaties. Hoofrapport en Bijlagenrapport. (The Position of the Pedestrian on Crossing Locations. Main report and appendices report).* De voetgangersvereniging (Dutch Pedestrians Association), The Hague, Netherlands.

Levelt, P.B.M. (1990). *Knipperend geel voor voetgangers: voorlopig en avontuur. (Flashing Yellow for Pedestrians: For the Time Being an Adventure).* R-90-23. SWOV, Leidschendam, Netherlands.

Levelt, P.B.M. (1992). *New Pedestrian Facilities: Technique, Observations and Opinions. The Dutch Experiment. Drive Project V1061: Pussycats.* R-92-55. SWOV, Leidschendam, Netherlands.

Levelt, P.B.M. (1994). *De opinie van voetgangers over de Maastrichtse opstelling. (Pedestrian Opinion on the Alternative "Maastricht" Crossing).* R-94-6. SWOV, Leidschendam, Netherlands.

Lourens, P.F., Molen, H.H. van der & Oude Egberink, H.J.H. (1986). *Driving Strategies Among Younger and Older Drivers When Encountering Children.* Pergamon Press plc, Headington Hill Hall, Oxford, UK.

Methorst, R. (1993). "Pedestrian Safety in the Next Decades." *Proceedings of the Conference on Strategic Highway Research Program (SHRP) and Traffic Safety on Two Continents.* The Hague, Netherlands.

Molen, H.H. van der & Linden L. van der (1987). "Pedestrian Injury Prevention." *Proceedings of the Canadian Multidisciplinary Road Safety Conference V* (p.186-198). Calgary, Alberta, Canada.

Oude Egberink, H.J.H. & Rothengatter, J.A. (1984). *Roodlichtnegatie van voetgangers op geregelde oversteekplaatsen. (Red Light Violation of Pedestrians at Signalized Crossing Places).* Rijksuniversiteit Groningen, Verkeerskundig Studiecentrum VSC, Haren, Netherlands.

Prikken, L.J.J. & Gerretsen, J.J. (1988). *Verkeersvoorzieningen voor mensen met een handicap. Handleiding en vragenlijst voor inventarisatie van verkeersruimte naar toegankelijkheid voor lichamelijk gehandicapten. (Traffic Provisions for Disabled Persons. Manual and Questionnaire for the Inventory of Traffic Provisions to the Accessibility for Disabled Persons).* Distributiecentrum DOP, The Hague, Netherlands.

Slop, M. & Minnen, J. van (1994). *Duurzaam-veilig voetgangers- en fietsverkeer. Een nadere uitwerking van het concept Duurzaam Veilig vanuit het perspectief van de voetganger en de fietser. (Sustainably Safe Pedestrian and Bicycle Traffic. A Further Elaboration of the Sustainable Safety Concept from the Perspective of the Pedestrian and the Cyclist).* R-94-67. SWOV, Leidschendam, Netherlands.

Spek, M. van der & Noyon, R. (1993). *Uitgeknikkerd, Opgehoepeld; Een onderzoek naar de bewegingsvrijheid van kinderen op straat. (A Survey on the Freedom of Movement of Children in Traffic).* Kinderen voorrang, The Hague, Netherlands.

Vis, A.A. & Kaal, I. (1993). *De veiligheid van 30 km/h-gebieden; een analyse van letselongevallen in 151 heringerichte gebieden in Nederlandse gemeenten. (Traffic Safety in 30-km/h Zones; an Analysis of Injury Accidents in 151 30-km/h Zones).* R-93.17. SWOV, Leidschendam, Netherlands.

Voetgangersvereniging (Dutch Pedestrians Association) (1988/1993). *Kinderen veilig naar school; een lange weg te gaan. (Survey on School Safety Zones).* The Hague, Netherlands.

VOR-ongevalsstatistieken 1980-1997. (Accident Records Registration Division of the Directorate-General of Public Works, 1980-1997). Heerlen, Netherlands.

Wolffelaar, P.C. van (1988). *Oudere verkeersdeelnemers: Verkeersproblemen en educatiedoelstellingen. (Elderly Traffic Participants: Traffic Problems and Educational Objectives).* Rijksuniversiteit Groningen, Verkeerskundig Studiecentrum VSC, Haren, Netherlands.

Wouters, P.I.J. (1991). *De veiligheid van oudere verkeersdeelnemers. (The Safety of Elderly Road Users).* R-91-77. SWOV, Leidschendam, Netherlands.

www.ingramcontent.com/pod-product-compliance
Lightning Source LLC
Chambersburg PA
CBHW081405170526
45166CB00010B/3212